YOUR KNOWLEDGE HAS VALUE

- We will publish your bachelor's and
  master's thesis, essays and papers

- Your own eBook and book -
  sold worldwide in all relevant shops

- Earn money with each sale

Upload your text at www.GRIN.com
and publish for free

Arjun Nelikanti

Colorectal Cancer MRI Image Segmentation Using Image Processing Techniques

GRIN Publishing

Bibliographic information published by the German National Library:

The German National Library lists this publication in the National Bibliography;
detailed bibliographic data are available on the Internet at http://dnb.dnb.de .

Imprint:

Copyright © 2014 GRIN Verlag GmbH
Print and binding: Books on Demand GmbH, Norderstedt Germany
ISBN: 978-3-656-87954-1

This book at GRIN:

http://www.grin.com/en/e-book/287641/colorectal-cancer-mri-image-segmentation-
using-image-processing-techniques

ACKNOWLEDGEMENT

The satisfaction that accompanies the successful completion of the task would be put incomplete without the mention of the people who made it possible, whose constant guidance and encouragement crown all the efforts with success.

I thankful to my guide **Mr. H. Venkateswara Reddy, Associate Professor, Head of the Department, Computer Science and Engineering** for his sustained inspiring guidance and cooperation throughout the process of this project. His wise counsel and suggestions were invaluable.

I would like to thank **Mr. H. Venkateswara Reddy, Associate Professor, Head of the Department, Computer Science and Engineering** for encouragement at various levels of my Project.

I would like to thank **Dr. B. Venkatesh, Professor, Dean of P.G Courses** for encouragement at various levels of my Project.

I show gratitude to **Dr. S. Sai Satyanarayana Reddy, Principal** for having provided all the facilities and support.

I avail this opportunity to express my deep sense of gratitude and hearty thanks to **Dr. Teegala Vijender Reddy,** Chairman and **Sri Teegala Upender Reddy,** Secretary of VCE, for providing congenial atmosphere and encouragement.

I express my deep sense of gratitude and thanks to all the **Teaching** and **Non-Teaching Staff** of my college who stood with me during the project and helped me to make it a successful venture.

I place highest regards to my **Parents, Friends** and **Well wishers** who helped a lot in making the report of this project.

N. Arjun

ABSTRACT

Colorectal cancer is the third most commonly diagnosed cancer and the second leading cause of cancer death in men and women. Magnetic resonance imaging (MRI) established itself as the primary method for detection and staging in patients with colorectal cancer. MRI images of Colorectal cancer are used to detect the area and mean values of tumor area and distance from tumor area to other parts. The thesis describes algorithms for preprocessing, clustering and post processing of MRI images. Implemented algorithm for preprocessing using image enhancement techniques, clustering is done using adaptive k-means algorithm and post processing using image processing techniques in MATLAB.

TABLE OF CONTENTS

Page No.

LIST OF TABLES

LIST OF FIGURES

ABBREVIATIONS

MRI	Magnetic Resonance Imaging
CT	Computed Tomography
PET	Positron Emission Tomography
CRM	Circumferential Resection Margin
TME	Total Mesorectal Excision
CLAHE	Using Contrast Limited Adaptive Histogram Equalization
PSNR	Peak-Signal-To-Noise-Ratio
AMBE	Absolute Mean Brightness Error
MSE	Mean Square Error

CHAPTER 1. INTRODUCTION

Cancer is a disease that begins in the cells of the body. In normal situations, the cells grow and divide as the body needs them. No more, no less. This orderly process is disturbed when new cells form that the body were not needed and old cells don't die when they should. These extra cells lump together to form a growth or tumor. One of the key problems in the treatment of cancer is the early detection of the disease. Often, cancer is detected in its later stages, when it has compromised the function of one or more vital organ systems and is widespread throughout the body. Methods for the early detection of cancer are of utmost importance and are an active area of current research.

Colorectal cancer is cancer that starts in the colon or the rectum. These cancers can also be referred to separately as colon cancer or rectal cancer, depending on where they start.

Colon cancer and rectal cancer have many features in common. They are discussed together in this document except for the section about treatment. Colorectal cancer is the third most common cancer diagnosed in both men and women in the United States. The American Cancer Society's estimates for the number of colorectal cancer cases in the United States for 2013 are 102,480 new cases of colon cancer and 40,340 new cases of rectal cancer. The death rate (the number of deaths per 100,000 people per year) from colorectal cancer has been dropping in both men and women for more than 20 years. Screening is also allowing more colorectal cancers to be found earlier when the disease is easier to cure. Regular colorectal cancer screening is one of the most powerful weapons for preventing colorectal cancer. Screening is the process of looking for cancer or pre-cancer in people who have no symptoms of the disease.

Cancer screening is the process of looking for cancer in people who have no symptoms of the disease. Several different tests can be used to screen for colorectal cancers. These tests can be divided into two broad groups:

- Tests that can find both colorectal polyps and cancer: These tests look at the structure of the colon itself to find any abnormal areas. This is done either with a scope inserted into the rectum or with special imaging tests. Polyps found before they become cancerous can be removed, so these tests may prevent colorectal cancer. This is why these tests are preferred if they are available and you are willing to have them.

- Tests that mainly find cancer: These test the stool (feces) for signs that cancer may be present. These tests are less invasive and easier to have done, but they are less likely to detect polyps.

After the initial detection of a cancerous growth, accurate diagnosis and staging of the disease are essential for the design of a treatment plan. This process is dependent on clinical testing and the observations of physicians. It is important for cancer patients and their families to understand the results given to them so that they can take an active role in the planning of the treatment protocol to be used. Magnetic resonance imaging (MRI) is a useful modality for the evaluation of rectal cancer, providing superior anatomic/pathologic visualization when compared with computed tomography (CT).

Preoperative MRI is useful for tissue characterization and tumor staging, which determines the surgical approach and need for neoadjuvant therapy. The stage describes the extent of the cancer in the body. It is based on how far the cancer has grown into the wall of the intestine, whether or not it has reached nearby structures, and whether or not it has spread to the lymph nodes or distant organs. The staging of a cancer is one of the most important factors in determining prognosis and treatment options. Important prognostic factors include the circumferential resection margin (CRM), T and N stages, and M extent of local invasion. The TNM system describes 3 key pieces of information:

- **T** describes how far the main (primary) tumor has grown into the wall of the intestine and whether it has grown into nearby areas.
- **N** describes the extent of spread to nearby (regional) lymph nodes. Lymph nodes are small bean-shaped collections of immune system cells that are important in fighting infections.
- **M** indicates whether the cancer has spread (metastasized) to other organs of the body. (Colorectal cancer can spread almost anywhere in the body, but the most common sites of spread are the liver and lungs.)

Numbers or letters appear after T, N, and M to provide more details about each of these factors. The numbers 0 through 4 indicate increasing severity.

1.1 OBJECTIVES

The objective of this real time processing of MRI image is to provide with information about the tumor. The inputs for this application are MRI images. The MRI images obtained may contain the noise or low contrast which is ought to be removed and enhanced. The objective of the thesis is to detect the tumor area and stage the colorectal cancer, the mean values and distance of tumor in MRI images of different cases are calculated for which image processing with MATLAB is utilized along with clustering techniques so as to analyze the input and generate information.

1.2 EXISTING SYSTEM

There are many preprocessing procedures are proposed earlier which differs from the proposed procedure and Computer-Aided Diagnostic (CAD) method for breast cancer image segmentation is done. And many systems are proposed for colorectal cancer detection by diagnosing colonic polyps at CT colonography and CT images.

1.3 PROPOSED SYSTEM

A CAD method that helps in detection and staging colorectal cancer is proposed. The proposed system uses MRI images as input. The raw input MRI image may not be suitable to process, therefore we need to preprocess before using them. Using 2D median filter, image sharpening and histogram equalization are certain image processing techniques which makes the image enhanced and suitable to get desired outputs. Adaptive k-means clustering algorithm used for image segmentation. The proposed system is implemented in MATLAB.

1.4 SCOPE:

This thesis would not only make the detection of cancer using clustering techniques but also used to stage the cancer. It represents a new technique in MRI image segmentation for detection of tumor. This thesis helps the radiologist in detection and staging the cancer.

1.5 ORGANIZATION OF THESIS

The rest of the thesis is organized as follows: Chapter II gives a glance to the literature review. Chapter III describes methodology. Chapter IV describes implementation. Chapter V illustrates results and discussions. Chapter VI ends up with conclusion of the paper.

CHAPTER 2. LITERATURE REVIEW

To determine the accuracy of MRI in the preoperative staging and planning of surgical management of rectal carcinoma. Rectal cancer constitutes about one-third of all gastrointestinal tract tumors. Because of its high recurrence rates reaching 30%, it is vitally important to accurately stage these tumors preoperatively, so that appropriate surgical resection can be undertaken. MRI is used to assist in staging, identifying patients who may benefit from preoperative chemotherapy–radiation therapy, and in surgical planning. [1] MRI of rectal cancer is accurate for prediction of tumor stage and the feasibility of sphincter-sparing surgery, which are the main factors affecting the outcome of surgery.

The [2] study was to develop a Hessian matrix–based computer-aided detection (CAD) algorithm for polyp detection on CT colonography (CTC) and to analyze its performance in a high-risk population. A Hessian matrix–based CAD algorithm for CTC has the potential to depict polyps larger than or equal to 6 mm with high sensitivity and an acceptable false-positive rate.

Imaging in rectal cancer helps in deciding the treatment and determining the prognosis. The newer techniques help in superior image resolution, and sometimes with functional qualities [3]. The most accurate method of rectal wall staging of rectal cancer is endorectal ultrasound and MRI but accurate staging of mesorectal fascia and lymph nodes is by phased array MRI.

The advances that have been made in the treatment of rectal cancer in recent years and that have considerably improved the prognosis of affected patients rely on differentiated tumor staging. [4]Despite its known limitations in T-staging, MRI is currently the only imaging modality that enables highly accurate evaluation of the topographic relationship between lateral tumor extent and the mesorectal fascia and to thus make a prediction about the CRM.

Image enhancement methods are utilized as pre- processing apparatus for further image dealing. the image enhancement in medical field is basically carried out in order to make things clearer and accurate that helps to effectively and precisely determine and deal with the problems and facts. [5] The survey shows many enhancement techniques.

The CLAHE method can be divided into steps [6] to achieve as following: The Medical image is divided into contextual regions which are continuous and non-overlapping. Each contextual region size is M×N. The histograms of each contextual regions are calculated and are clipped.

Magnetic resonance imaging is a valuable tool for the planning of treatment in rectal cancer [7]. It is the single most important technique in the local staging of patients with rectal cancer.

Cancer staging represents the operational basis for choosing the most appropriate therapy and for evaluating the efficacy of different therapeutic methods; it is an essential component of patient care, cancer research, and control activities, even in light of the impressive progress that has been attained in the fields of clinical strategies and molecular medicine. The [8] TNM system is subjected to continuous updating through an ongoing expert review of existing data.

The rectum has curvatures both in the right-left direction and in the antero-posterior direction. In addition, the rectum, as the name implies, especially when filled with a tumor has a spherical rather than a cylindrical shape and is thus more difficult to image perpendicular to its wall than if cylindrical. Finally, the pelvic floor is formed like a funnel necessitating different image planes than the three traditional orthogonal planes for adequate assessment. It can be difficult for non-specialized MRI technicians to find a rectal tumor and to anticipate and plan the right imaging planes.

Surrounding the rectum, there is a layer of fat, the perirectal or the mesorectal fat. There is less fat anterior and caudal to the rectum than to the other sides. The amount of perirectal fat is larger in men and correlates with the visceral compartment area, but not the body's cross-sectional area, body mass index or age. Although the total amount of fat has not been shown to influence the accuracy of tumor staging, it is conceivable that the small distance between structures ventral to the rectum can make analysis of tumor growth assessment ventrally difficult. The perirectal fat is often referred to as mesorectum. The fascia is fused to the mesorectal fascia as it concerns MRI. The mesorectal or perirectal visceral fascia has gained great importance as it pertains to rectal cancer surgery when total mesorectal excision (TME) was introduced. Tumors located close to this fascia are considered to threaten the fascia. What constitutes a safe distance is, however, no ascertained. A distance <1 mm is definitely associated with risk of involvement of the surgical circumferential resection margin (CRM). Distance <2-5 mm has also been suggested as indicative of a threatened margin with a greater risk of recurrence.

CHAPTER 3. METHODOLOGY

The methodology in this thesis used is based on different platforms like data mining of which clustering technique is as the basic paradigm and image processing.

3.1 CLUSTERING

Clustering can be considered the most important unsupervised learning problem; so, as every other problem of this kind, it deals with finding a structure in a collection of unlabeled data[9].

A loose definition of clustering could be "the process of organizing objects into groups whose members are similar in some way".

A cluster is therefore a collection of objects which are "similar" between them and are "dissimilar" to the objects belonging to other clusters.

We can show this with a simple graphical example:

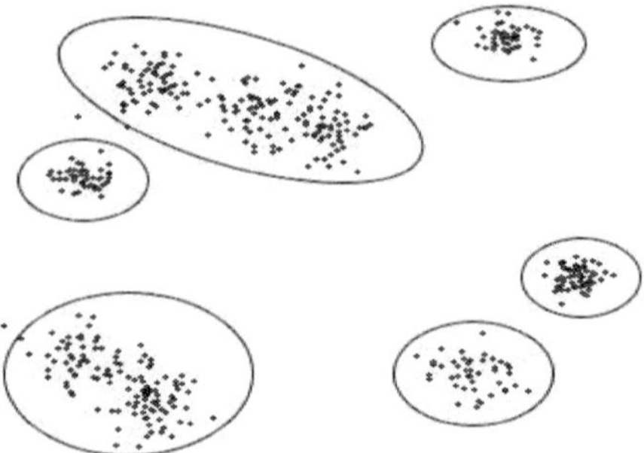

Figure 3.1.1 Clustering Objects.

In this case we easily identify the 4 clusters into which the data can be divided; the similarity criterion is distance: two or more objects belong to the same cluster if they are "close" according to a given distance (in this case geometrical distance). This is called *distance-based clustering*. Another kind of clustering is *conceptual clustering*: two or more objects belong to the same cluster if this one defines a concept *common* to all that objects. In other words, objects are grouped according to their fit to descriptive concepts, not according to simple similarity measures.

The Goals of Clustering

So, the goal of clustering is to determine the intrinsic grouping in a set of unlabeled data. But how to decide what constitutes a good clustering? It can be shown that there is no absolute "best" criterion which would be independent of the final aim of the clustering. Consequently, it is the user which must supply this criterion, in such a way that the result of the clustering will suit their needs.

For instance, we could be interested in finding representatives for homogeneous groups (*data reduction*), in finding "natural clusters" and describe their unknown properties (*"natural" data types*), in finding useful and suitable groupings (*"useful" data classes*) or in finding unusual data objects (*outlier detection*).

Possible Applications

Clustering algorithms can be applied in many fields, for instance:

Marketing: finding groups of customers with similar behavior given a large database of customer data containing their properties and past buying records;

Biology: classification of plants and animals given their features;

Libraries: book ordering;

Insurance: identifying groups of motor insurance policy holders with a high average claim cost; identifying frauds;

City-planning: identifying groups of houses according to their house type, value and geographical location;

Earthquake studies: clustering observed earthquake epicenters to identify dangerous zones;

WWW: document classification; clustering weblog data to discover groups of similar access patterns.

Problems

There are a number of problems with clustering. Among them: current clustering techniques do not address all the requirements adequately (and concurrently); dealing with large number of dimensions and large number of data items can be problematic because of time complexity; the effectiveness of the method depends on the definition of "distance" (for distance-based clustering); if an *obvious* distance measure doesn't exist we must "define" it, which is not always easy, especially in multi-dimensional spaces; the result of the clustering algorithm (that in many cases can be arbitrary itself) can be interpreted in different ways.

Clustering Algorithms

Classification

Clustering algorithms may be classified as listed below:

- Exclusive Clustering

- Overlapping Clustering

- Hierarchical Clustering

- Probabilistic Clustering

In the first case data are grouped in an exclusive way, so that if a certain datum belongs to a definite cluster then it could not be included in another cluster. On the contrary the second type, the overlapping clustering, uses fuzzy sets to cluster data, so that each point may belong to two or more clusters with different degrees of membership. In this case, data will be associated to an appropriate membership value. Instead, a hierarchical clustering algorithm is based on the union between the two nearest clusters. The beginning condition is realized by setting every datum as a cluster. After a few iterations it reaches the final clusters wanted. Finally, the last kind of clustering uses a completely probabilistic approach.

The k-means algorithms starts with the selection of K elements from the input data set. The K elements form the seeds of clusters and are randomly selected. The properties of each element also form the properties of the cluster that is constituted by the element. The algorithm is based on the ability to compute distance between a given element and a cluster. This function is also used to compute distance between two elements. An important consideration for this function is that it should be able to account for the distance based on properties that have been normalized so that the distance is not dominated by one property or some property is not ignored in the computation of distance. In most cases, the Euclidean distance may be sufficient. For example, in the case of spectral data given by n-dimensions, the distance between two data elements

$E_1 = \{E_{11}, E_{12}, \ldots ,E_1\}$ and $E_2 = \{E_{21}, E_{22}, \ldots ,E2\}$

is given by

$$((E_{11} - E_{12})^2 + (E_{12} - E_{22})^2 + \ldots + (E_{1n} - E_{2n})^2)^{1/2}$$

It should be pointed out that for performance reasons, the square root function may be dropped. In other cases, we may have to modify the distance function. Such cases can be exemplified by data where one dimension is scaled different compared to other dimensions, or where properties may be required to have different weights during comparison. With the distance function, the algorithm proceeds as follows:

Compute the distance of each cluster from every other cluster. This distance is stored in a 2D array as a triangular matrix. We also note down the minimum distance d_{min} between any two clusters Cm1 and Cm2 as well as the identification of these two closest clusters.

For each unclustered element E_i, compute the distance of E_i from each cluster. For assignment of this element to a cluster, there can be three cases as follows:

1. If the distance of the element from a cluster is 0, assign the element to that cluster, and start working with the next element.

2. If the distance of the element from a cluster is less than the distance d_{min}, assign this element to its closest cluster. As a result of this assignment, the cluster representation, or centroid, may change. The centroid is recomputed as an average of properties of all elements in the cluster. In addition, we recomputed the distance of the affected cluster from every other cluster, as well as the minimum distance between any two clusters and the two clusters that are closest to each other.

3. The last case occurs when the distance d_{min} is less than the distance of the element from the nearest cluster. In this case, we select the two closest clusters C_{m1} and C_{m2} , and merge C_{m2} into C_{m1} . Also, we destroy the cluster C_{m2} by removing all the elements from the cluster and by deleting its representation. Then, we add the new element into this now empty cluster, effectively creating a new cluster. The distances between all clusters are recomputed and the two closest clusters identified again.

The above three steps are repeated until all the elements have been clustered. There is a possibility that the algorithm identifies a number of singletons or single-element clusters if the distance of some elements is large from other elements. These elements are known as outliers and can be accounted for by looking for clusters with an extremely small number of elements and removing those elements from clustering consideration, or handled as exceptions.

The adaptive K-means clustering algorithm as follows:

- Initialize seed point as mean of the input image and counter for each iteration.
- Calculating distance between seed and gray values to set bandwidth for cluster center.
- Check values are in selected bandwidth or not and update mean.
- Remove values which have assigned to a cluster and store center of cluster.
- Update seed and sort centers.
- Finding out difference between two consecutive centers.
- Discard cluster centers less than distance and make a clustered images using these centers.

3.2 INPUT/OUTPUT:

IMAGE:

An image is a two-dimensional picture, which has a similar appearance to some subject usually a physical object or a person. Image is a two-dimensional, such as a photograph, screen display, and as well as a three-dimensional, such as a statue. They may be captured by optical devices—such as cameras, mirrors, lenses, telescopes, microscopes, etc. and natural objects and phenomena, such as the human eye or water surfaces.

The word image is also used in the broader sense of any two-dimensional figure such as a map, a graph, a pie chart, or an abstract painting. In this wider sense, images can also be rendered manually, such as by drawing, painting, carving, rendered automatically by printing or computer graphics technology, or developed by a combination of methods, especially in a pseudo-photograph.

An image is a rectangular grid of pixels. It has a definite height and a definite width counted in pixels. Each pixel is square and has a fixed size on a given display. However different computer monitors may use different sized pixels. The pixels that constitute an image are ordered as a grid (columns and rows); each pixel consists of numbers representing magnitudes of brightness and color.

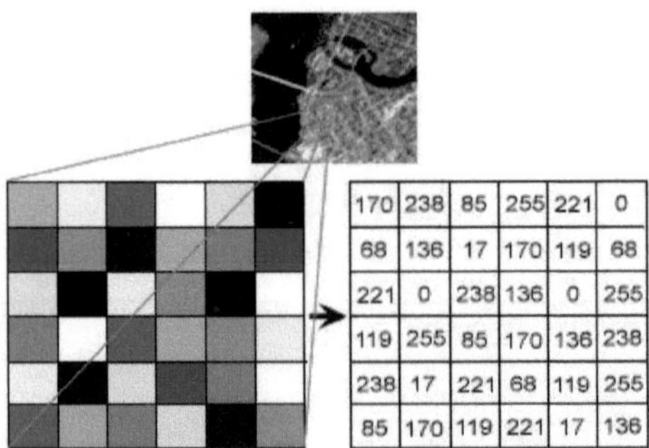

Figure 3.2.1 Pixels In Particular Region

Each pixel has a color. The color is a 32-bit integer. The first eight bits determine the redness of the pixel, the next eight bits the greenness, the next eight bits the blueness, and the remaining eight bits the transparency of the pixel.

IMAGE FILE SIZES:

Image file size is expressed as the number of bytes that increases with the number of pixels composing an image, and the color depth of the pixels. The greater the number of rows and columns, the greater the image resolution, and the larger the file. Also, each pixel of an image increases in size when its color depth increases, an 8-bit pixel (1 byte) stores 256 colors, a 24-bit pixel (3 bytes) stores 16 million colors, the latter known as true color. Image compression uses algorithms to decrease the size of a file.

High resolution cameras produce large image files, ranging from hundreds of kilobytes to megabytes, per the camera's resolution and the image-storage format capacity. High resolution digital cameras record 12 megapixel (1MP = 1,000,000 pixels / 1 million) images, or more, in true color.

For example, an image recorded by a 12 MP camera; since each pixel uses 3 bytes to record true color, the uncompressed image would occupy 36,000,000 bytes of memory, a great amount of digital storage for one image, given that cameras must record and store many images to be practical.

IMAGE FILE FORMATS:

Image file formats are standardized means of organizing and storing images. This entry is about digital image formats used to store photographic and other images. Image files are composed of either pixel or vector (geometric) data that are rasterized to pixels when displayed (with few exceptions) in a vector graphic display. Including proprietary types, there are hundreds of image file types. The PNG, JPEG, and GIF formats are most often used to display images on the Internet.

JPEG (Joint Photographic Experts Group) is a lossy compression method; JPEG-compressed images are usually stored in the **JFIF** (JPEG File Interchange Format) file format. The JPEG/JFIF filename extension is **JPG** or **JPEG**. Nearly every digital camera can save images in the JPEG/JFIF format, which supports 8-bit grayscale images and 24-bit color images (8 bits each for red, green, and blue). JPEG applies lossy compression to images, which can result in a significant reduction of the file size. Applications can determine the degree of compression to apply, and the amount of compression affects the visual quality of the result. When not too great, the compression does not noticeably affect or detract from the image's quality, but JPEG files suffer generational degradation when repeatedly edited and saved. (JPEG also provides lossless image storage, but the lossless version is not widely supported).

GIF (Graphics Interchange Format) is limited to an 8-bit palette, or 256 colors. This makes the GIF format suitable for storing graphics with relatively few colors such as simple diagrams, shapes, logos and cartoon style images. The GIF format supports animation and is still widely used to provide image animation effects. Its LZW lossless compression is more effective when large areas have a single color, and less effective for photographic or dithered images.

The **PNG** (Portable Network Graphics) file format was created as a free, open-source alternative to GIF. The PNG file format supports 8 bit paletted images (with optional transparency for all palette colors) and 24 bit truecolor (16 million colors) or 48 bit truecolor with and without alpha channel - while GIF supports only 256 colors and a single transparent color.

There are different type of format used in medical field called Digital Imaging and Communications in Medicine (DICOM).

3.3 IMAGE PROCESSING

Digital image processing [10] methods were introduced in 1920, when people were interested in transmitting picture information across the Atlantic Ocean. The time taken to transmit one image of size 256×256 was about a week. The pictures were encoded using specialized printing equipment and were transmitted through the submarine cable. At the receiving end, the coded pictures were reconstructed. The reconstructed pictures were not up to the expected visual quality and the contents could not be interpreted due to the interference. Hence, the scientists and engineers who were involved in the transmission of picture information started devising various techniques to improve the visual quality of the pictures. This was the starting point for the introduction of the image processing methods. To improve the speed of transmission the Bart lane cable was introduced and it reduced the *transmission time* of the picture information from 1 week to less than three hours. In the early stages, attempts to improve the visual quality of the received image were related to the selection of printing procedures and distribution of brightness levels. During the 1920s the coding of images involved five distinct brightness levels. In the year 1929, the number of brightness levels was increased to 15 and this improved the visual quality of the images received.

The use of digital computers for improving the quality of the images received from space probe began at the Jet Propulsion Laboratory in the year 1964. From 1964 until today the field of digital image processing has grown vigorously. Today, digital image processing techniques are used to solve a variety of problems. These techniques are used in two major application areas and they are

1. Improvement of pictorial information for human interpretation
2. Processing of scene of data for autonomous machine perception.

The following paragraphs give some of the application areas where image processing techniques are capable of enhancing pictorial information for human interpretation. In medicine, the digital image processing techniques are used to enhance the contrast or transform the intensity levels into color for easier interpretation of X-rays and other bio-medical images. The geographers will make use of the available image processing techniques to enhance the pollution patterns from aerial and satellite imagery.

Image enhancing techniques can be used to process degraded images of unrecoverable objects or experimental results too expensive to duplicate. In archeology, image processing techniques have successfully restored blurred pictures that were the only available records of rare artifacts lost or damaged after being photographed. The following examples illustrate the digital image processing techniques dealing with problems in machine perception. The character recognition, industrial machine vision for product assembly and inspection, fingerprint processing, and weather prediction are some of the problems in machine perception that utilize the image processing techniques.

Steps In Image Processing

The various steps required for any digital image processing applications are listed below:

1. Image grabbing or acquisition
2. Preprocessing
3. Segmentation
4. Representation and feature extraction
5. Recognition and interpretation.

It is more appropriate to explain the various steps in digital image processing with an application like mechanical components classification system. Let us consider an industrial application where the production department is involved in the manufacturing of certain mechanical components like bolts, nuts, and washers. Periodically, each one of these components must be sent to the stores via a conveyor belt and these components are dropped in the respective bins in the store room.

In the image acquisition step using the suitable camera, the image of the component is acquired and then subjected to digitization. The camera used to acquire the image can be a monochrome or color TV camera which is capable of producing images at the rate of 25 images per sec.

The second step deals with the preprocessing of the acquired image. The key function of preprocessing is to improve the image such that it increases the chances for success of other processes. In this application, the preprocessing techniques are used for enhancing the contrast of the image, removal of noise and isolating the objects of interest in the image.

The next step deals with segmentation—a process in which the given input image is partitioned into its constituent parts or objects. The key role of segmentation in the mechanical component classification is to extract the boundary of the object from the background. The output of the segmentation stage usually consists of either boundary of the region or all the parts in the region itself. The boundary representation is appropriate when the focus is on the external shape and regional representation is appropriate when the focus is on the internal property such as texture. The application considered here needs the boundary representation to distinguish the various components such as nuts, bolts, and washers.

In the representation step the data obtained from the segmentation step must be properly transformed into a suitable form for further computer processing. The feature selection deals with extracting salient features from the object representation in order to distinguish one class of objects from another. In terms of component recognition the features such as the inner and the outer diameter of the washer, the length of the bolt, and the length of the sides of the nut are extracted to differentiate one component from another.

The last step is the recognition process that assigns a label to an object based on the information provided by the features selection. Interpretation is nothing but assigning meaning to the recognized object. We have not yet discussed about the prior knowledge or the interaction between the knowledge base and the processing modules. Knowledge about the problem domain is coded into the image processing system in the form of knowledge database. This knowledge is as simple as describing the regions of the image where the information of interest is located. Each module will interact with the knowledge base to decide about the appropriate technique for the right application. For example, if the acquired image contains spike-like noise the preprocessing module interacts with the knowledge base to select an appropriate smoothing filter-like median filter to remove the noise.

Building Blocks Of A Digital Image Processing System

The major building blocks of a digital image processing system are as follows:

1. Acquisition
2. Storage
3. Processing
4. Display and communication interface.

(1) Image Acquisition

In order to acquire a digital image, a physical device sensitive to a band in the electromagnetic energy spectrum is required. This device converts the light (X-rays, ultraviolet, visible, or infrared) information into corresponding electrical signal. In order to convert this electrical signal into digital signal another device called digitizer is employed.

(2) Storage

There are three different types of digital storage that are available for digital image processing applications. The first type of storage or memory used during processing is called short-term storage. For frequent retrieval of the images the second type of storage called online storage *is* employed. The third type of memory is called archival storage, characterized by infrequent access. One way of providing short-term memory is by using the main memory of the computer. Another way of implementing the short-term memory is using specialized boards called frame buffers. When the images are stored in frame buffers, they can be accessed rapidly at the rate of 30 images per second. The images in the frame buffers allow operations such as instantaneous image zoom, vertical shift, and horizontal shift. Frame buffer cards are available to accommodate as many images as 32 (32 MB).

The online memory generally uses Winchester disks of capacity 1 GB. In the recent years magneto optical storage became popular. It uses a laser and specialized material to achieve few gigabytes of storage on optical disk. Since the online storage is used for frequent access of data, the magnetic tapes are not used. For large online storage capacity 32 to 100 optical disks are in a box and this arrangement is called Jukebox.

The archival storage is usually larger in size and is used for infrequent access. High-density magnetic tapes and Write–Once–Read–Many (WORM) optical disk is used for realizing the archival memory. Magnetic tapes are capable of storing 6.4KB per inch of image data and therefore to store one megabyte of image requires 13 ft of tape. WORM disks with a capacity to store 6GB on 12? disk and 10GB on 14? disk are commonly available. The lifetime of the magnetic tape is only 7 years, whereas for WORM disk it is more than 30 years.

(3)Processing

The processing of images need a specialized hardware consisting of a high-speed processor. This processor is totally different from the conventional processor available in a computer. The processor and the associated hardware is realized in the form of a card called image processor card. The processor in the cards is capable of processing the data of different word size. For example, the image processor card IP-8 is capable of processing word size of 8 bits. The image processor card usually consists of a digitizer, a frame buffer, the arithmetic and logical unit (ALU), and the display module.

The digitizer is nothing but an analog to digital converter to convert the electrical signal corresponding to the intensities of the optical image into a digital image. There may be one or more frame buffers for fast access to image data during processing. The ALU is capable of performing the arithmetic and logical operations at frame rate. Suitable software comes along with the image processor card to realize various image processing techniques/algorithms.

(4)Display and Communication Interface

Black and white and color monitors are used as display devices in the image processing system. These monitors are driven by the output signals from the display module, which is available in the image processor card. The signals of the display module can also be given to the recording device that produces the hard copy of the image being viewed on the monitor screen. The other display devices include dot matrix printer and laser printer. The image display devices are useful for low-resolution image processing works.

The communication interface is quite useful to establish communication between image processing systems and remote computers. Suitable hardware and software are available for this purpose. Different types of communication channels or media are available for extension of image data. For example, a telephone line can be used to transmit an image at a maximum rate of 9600 bits/sec. Fiber optic links, microwave links, and satellite links are much faster and cost considerably more.

Image enhancement is the improvement of digital image quality (wanted e.g. for visual inspection or for machine analysis), without knowledge about the source of degradation. If the source of degradation is known, one calls the process image restoration. Both are iconical processes, viz. input and output are images.

Many different, often elementary and heuristic methods are used to improve images in some sense. The problem is, of course, not well defined, as there is no objective measure for image quality. Here, we discuss a few recipes that have shown to be useful both for the human observer and/or for machine recognition. These methods are very problem-oriented: a method that works fine in one case may be completely inadequate for another problem.

Apart from geometrical transformations some preliminary grey level adjustments may be indicated, to take into account imperfections in the acquisition system. This can be done pixel by pixel, calibrating with the output of an image with constant brightness. Frequently space-invariant grey value transformations are also done for contrast stretching, range compression, etc. The critical distribution is the relative frequency of each grey value, the grey value histogram. Examples of simple grey level transformations in this domain are:

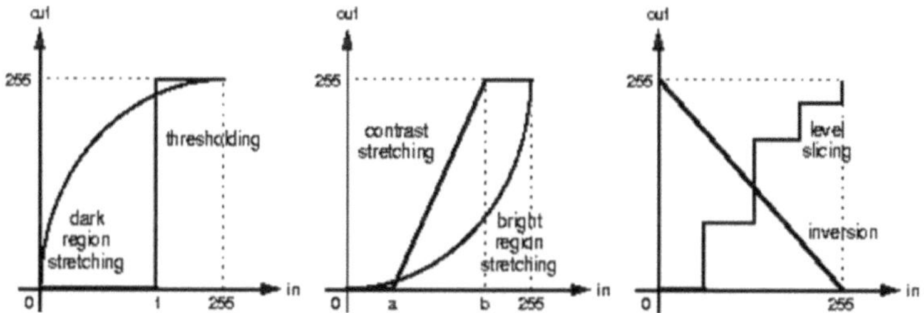

Figure 3.2.2 Gray Level Transformations.

Greyvalues can also be modified such that their histogram has any desired shape, e.g flat (every greyvalue has the same probability).

Physiological experiments have shown that very small changes in luminance are recognized by the human visual system in regions of continuous grey value, and not at all seen in regions of some discontinuities. Therefore, a design goal for image enhancement often is to smooth images in more uniform regions, but to preserve edges. On the other hand, it has also been shown that somehow degraded images with enhancement of certain features, e.g. edges, can simplify image interpretation both for a human observer and for machine recognition. A second design goal, therefore, is image sharpening. All these operations need neighbourhood processing, viz. the output pixel is a function of some neighbourhood of the input pixels.

These operations could be performed using linear operations in either the frequency or the spatial domain. We could, e.g. design, in the frequency domain, one-dimensional low or high pass filters (→ Filtering), and transform them to the two-dimensional case. Unfortunately, linear filter operations do not really satisfy the above two design goals; in this book, we limit ourselves to discussing separately only (and superficially) Smoothing and Sharpening.

Apart from point and neighbourhood processing, there are also global processing techniques, i.e. methods where every pixel depends on all pixels of the whole image. Histogram methods are usually global, but they can also be used in a neighbourhood.

3.4 IMAGE SEGMENTATION

Segmentation of images can proceed on three different ways,

- Manually
- Automatically
- Semi automatically

Manual Segmentation

The pixels belonging to the same intensity range could manually be pointed out, but clearly this is a very time consuming method if the image is large. A better choice would be to mark the contours of the objects. This could be done discrete from the keyboard, giving high accuracy, but low speed, or it could be done with the mouse with higher speed but less accuracy. The manual techniques all have in common the amount of time spent in tracing the objects, and human resources are expensive. Tracing algorithms can also make use of geometrical figures like ellipses to approximate the boundaries of the objects. This has been done a lot for medical purposes, but the approximations may not be very good.

Automatic Segmentation

Fully automatic segmentation is difficult to implement due to the high complexity and variation of images. Most algorithms need some a priori information to carry out the segmentation, and for a method to be automatic, this a priori information must be available to the computer. The needed apriori information could for instance be noise level and probability of the objects having a special distribution.

Semiautomatic Segmentation

Semiautomatic segmentation combines the benefits of both manual and automatic segmentation. By giving some initial information about the structures, we can proceed with automatic methods. Further clustering, boundary tracking thresholding is one of the techniques used for image segmentation.

MATLAB

MATLAB is widely used in all areas of applied mathematics, in education and research at universities, and in the industry. MATLAB stands for MATrix LABoratory and the software is built up around vectors and matrices. This makes the software particularly useful for linear algebra but MATLAB is also a great tool for solving algebraic and differential equations and for numerical integration. MATLAB has powerful graphic tools and can produce nice pictures in both 2D and 3D. It is also a programming language, and is one of the easiest programming languages for writing mathematical programs. MATLAB also has some tool boxes useful for signal processing, image processing, optimization, etc.

How to start MATLAB

Mac: Double-click on the icon for MATLAB.

PC: Choose the submenu "Programs" from the "Start" menu. From the "Programs" menu, open the "MATLAB" submenu. From the "MATLAB" submenu, choose "MATLAB".

Unix: At the prompt, type matlab.

You can quit MATLAB by typing exit in the command window.

Environment

The MATLAB environment (on most computer systems) consists of menus, buttons and a writing area similar to an ordinary word processor. There are plenty of help functions that you are encouraged to use. The writing area that you will see when you start MATLAB, is called the *command window*. In this window you give the commands to MATLAB. For example, when you want to run a program you have written for MATLAB you start the program in the command window by typing its name at the prompt. The command window is also useful if you just want to use MATLAB as a scientific calculator or as a graphing tool. If you write longer programs, you will find it more convenient to write the program code in a separate window, and then run it in the command window.

In the command window you will see a prompt that looks like >> . You type your commands immediately after this prompt. Once you have typed the command you wish MATLAB to perform, press <enter>. If you want to *interrupt a command* that MATLAB is running, type <ctrl> + <c>.

The commands you type in the command window are stored by MATLAB and can be viewed in the *Command History* window. To repeat a command you have already used, you can simply double-click on the command in the history window, or use the <up arrow> at the command prompt to iterate through the commands you have used until you reach the command you desire to repeat.

Matrix operations

An important thing to remember is that since MATLAB is matrix-based, the multiplication operator "*" denotes matrix multiplication. Therefore, A*B is not the same as multiplying each of the elements of A times the elements of B. However, you'll probably find that at some point you want to do element-wise operations (array operations). In MATLAB you denote an array operator by playing a period in front of the operator. The difference between "*" and ".*" is demonstrated in this example:

>> A = [1 1 1; 2 2 2; 3 3 3];

>> B = ones(3,3);

>> A*B

ans =

3 3 3

6 6 6

9 9 9

>> A.*B

ans =

1 1 1

2 2 2

3 3 3

Other than the bit about matrix vs. array multiplication, the basic arithmetic operators in MATLAB work pretty much as you'd expect. You can add (+), subtract (-), multiply (*), divide (/), and raise to some power (^). MATLAB provides many useful functions for working with matrices. It also has many scalar functions that will work element-wise on matrices (e.g., the function sqrt(x) will take the square root of each element of the matrix x).

Image processing toolbox is included in MATLAB [11]. MATLAB is a high-level language and interactive environment for numerical computation, visualization, and programming. Using MATLAB, you can analyze data, develop algorithms, and create models and applications. The language, tools, and built-in math functions enable you to explore multiple approaches and reach a solution faster than with spreadsheets or traditional programming languages, such as C/C++ or Java. One can use MATLAB for a range of applications, including signal processing and communications, image and video processing, control systems, test and measurement, computational finance, and computational biology. More than a million engineers and scientists in industry and academia use MATLAB, the language of technical computing.

Image Processing Toolbox provides a comprehensive set of reference-standard algorithms, functions, and apps for image processing, analysis, visualization, and algorithm development. You can perform image enhancement, image deblurring, feature detection, noise reduction, image segmentation, geometric transformations, and image registration. Many toolbox functions are multithreaded to take advantage of multicore and multiprocessor computers.

Image Processing Toolbox supports a diverse set of image types, including high dynamic range, giga pixel resolution, embedded ICC profile, and tomographic. Visualization functions let you explore an image, examine a region of pixels, adjust the contrast, create contours or histograms, and manipulate regions of interest (ROIs). With toolbox algorithms you can restore degraded images, detect and measure features, analyze shapes and textures, and adjust colour balance.

CHAPTER 4. IMPLEMENTATION

4.1 PROPOSED METHOD

The proposed method as follows:

> **Read input image :**

i.   The input image is in Joint Photographic Experts Group (JPEG) format taken from DICOM file.

ii. Now the input image is converted into grayscale image.

> **Preprocessing**

Steps in preprocessing as follows:

i.   Performing *median filtering* on the grayscale image :

In median filtering [12], the neighboring pixels are ranked according to brightness (intensity) and the median value becomes the new value for the central pixel.

Median filters can do an excellent job of rejecting certain types of noise, in particular, "shot" or impulse noise in which some individual pixels have extreme values.

122	123	134	145	125
114	**124**	**126**	**127**	129
117	**120**	**150**	**125**	133
111	**115**	**119**	**123**	144
110	113	140	130	120

Neighbouhood values:

115, 119, 120, 123, 124,

125, 126, 127, 150

Median value: 124

Figure 4.1.1 Median Filter 3x3

ii. Performing *unsharp masking* :

Sharpness is actually the contrast between different colors. A quick transition from black to white looks sharp. A gradual transition from black to gray to white looks blurry. Sharpening images increases the contrast along the edges where different colors meet. The unsharp masking technique comes from a publishing industry process in which an image is sharpened by subtracting a blurred (unsharp) version of the image from itself. Do not be confused by the name of this filter: an unsharp filter is an operator used to sharpen an image.

The unsharp masking algorithm can be described by the equation: $v = y + \gamma (x - y)$ where x is the input image, y is the result of a linear low-pass filter, and the gain($\gamma > 0$) is a real scaling factor. The detail signal $d = (x - y)$ is usually amplified to increase the sharpness.

iii.   Performing *Contrast Limited Adaptive Histogram Equalization (CLAHE)*:

CLAHE is an image processing technique used to improve contrast in images [13]. It differs from ordinary histogram equalization in the respect that the adaptive method computes several histograms, each corresponding to a distinct section of the image, and uses them to redistribute the lightness values of the image. It is therefore suitable for improving the local contrast of an image and bringing out more detail.

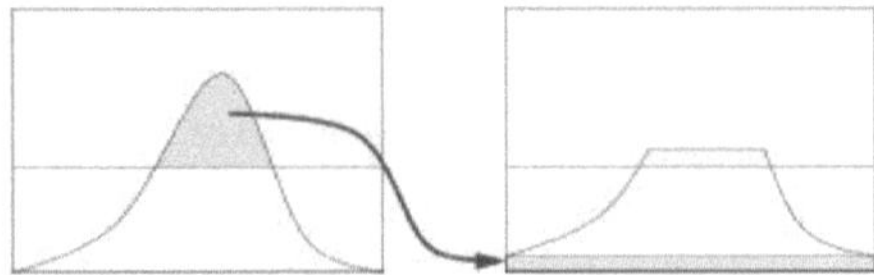

Figure 4.1.2 CLAHE Distribution.

The redistribution will push some bins over the clip limit again (region shaded green in the figure), resulting in an effective clip limit that is larger than the prescribed limit and the exact value of which depends on the image. If this is undesirable, the redistribution procedure can be repeated recursively until the excess is negligible.

A. Validation of enhanced image quality:

Calculate the PSNR value in order to check for the visual quality of the enhanced image.

$$PSNR = 10 \log_{10} \frac{255^2}{MSE}$$
(1)

where Mean Square Error (MSE) stands for the mean squared difference between the original image and the enhanced image. The mathematical definition for MSE is:

$$MSE = \left(\frac{1}{M \times N}\right) \sum_{i=1}^{M} \sum_{j=1}^{N} (a_{ij} - b_{ij})^2 \tag{2}$$

In Eq(2), a_{ij} means the pixel value at position (i, j) in the cover-image and b_{ij} means the pixel value at the same position in the corresponding enhanced image.

The calculated PSNR usually adopts dB value for quality judgment. The larger PSNR is, the higher the image quality is (which means there is only little difference between the original image and the enhanced image). On the contrary, a small dB value of PSNR means there is great distortion.

Absolute Mean Brightness Error (AMBE) [14] which is the difference between original and enhanced image and is given as:

$$AMBE = |\, E(x) - E(y)\,|. \tag{3}$$

where E(x) is average intensity of original image and E(y) is average intensity of enhanced image.

> **Clustering**

A popular technique for clustering is based on K-means. The image after enhancement is used to find clusters using Adaptive K-Means Clustering [15-16]. The algorithm starts with random selection of initial seeds from the image, these seed properties also form the properties of cluster.

The algorithm is as follows:

- Initialize seed point(mean).
- Find distance between seed and gray value. Find bandwidth for cluster center.
- Check values are in selected bandwidth or not. Update mean.
- Remove values which have assigned to a cluster. Store center of cluster.
- Update seed.
- Sort centers.
- Findout minimum distance between two cluster centers. Discard cluster centers less than distance.
- Make a clustered image using these centers. Find distance between center and pixel value.
- Choose cluster index of minimum distance. Reshape the labelled index vector.

This is the algorithm used for clustering the image after preprocessing.

➢ **Post processing**

The clustered image that is close to the required data is selected and processed further.

i. The binary image containing objects are returned by the function bwselect and boundaries are drawn for the selected.

ii. And the properties such as area and mean values are calculated for the objects selected using region properties.

iii. Area in mm^2 is calculated by the following

$$\text{Area} = ((pr)^{1\backslash2}) * 0.264. \tag{4}$$

where *pr* is number of pixels.

iv. Using distance line function we draw a line by default on the figure and used to calculate the distance between cancer boundary and the other parts in the image. The Distance tool is a draggable, resizable line, superimposed on an axes, that measures the distance between the two endpoints of the line. The Distance tool displays the distance in a text label superimposed over the line. The tools specifies the distance in data units determined by the XData and YData properties, which is pixels, by default.

v. The distance line measures the distance between the two endpoints of the line. The distance in a text label superimposed over the line specifies the distance in pixels. The distance in unit pixels is converted into centimeters by the following equation

$$dcm = dpix * 0.02645. \tag{5}$$

where *dpix* is distance in pixel and *dcm* is distance in centimeters.

CHAPTER 5. RESULTS AND DISCUSSION

The following steps are followed in order to set the satellite image as input, perform image segmentation using adaptive K-means clustering

5.1 Reading Input Image

Read in Image3.jpg, which is a MRI image consisting of different intensity. The image is converted into gray scale, and is segmented using clustering technique by which the required texture is taken into consideration leaving the rest as noise within the said image.

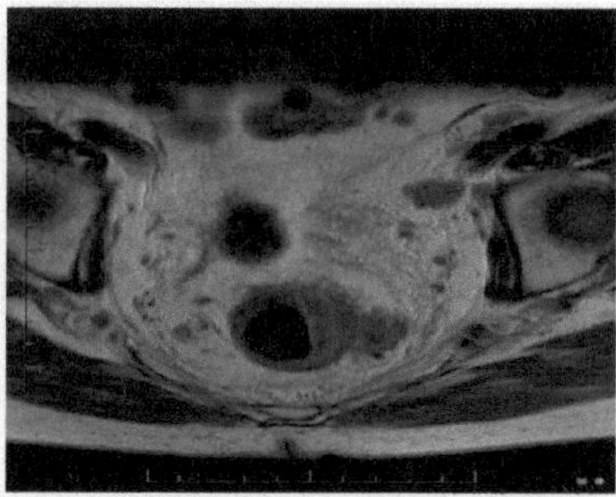

Figure 5.1.1 Original Image.

5.2 Pre processing

This step performs image enhancement by the applying median filter. Median filter replaces all the image pixels in the same time with the median of image pixel values in predefined (3-by-3) neighborhood of the given pixel. Now the filtered image is sharpened by using unsharp masking.

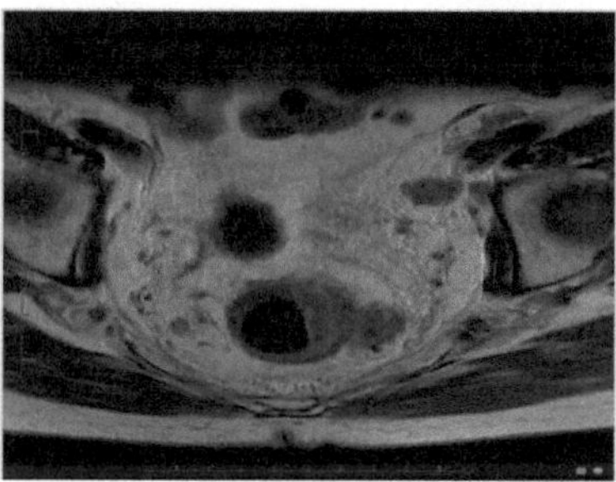

Figure 5.2.1 Image After Sharpening.

The final step in preprocessing is equalizing the contrast of the image. Using CLAHE equalization of the contrast and clarity of the sharpened image is obtained.

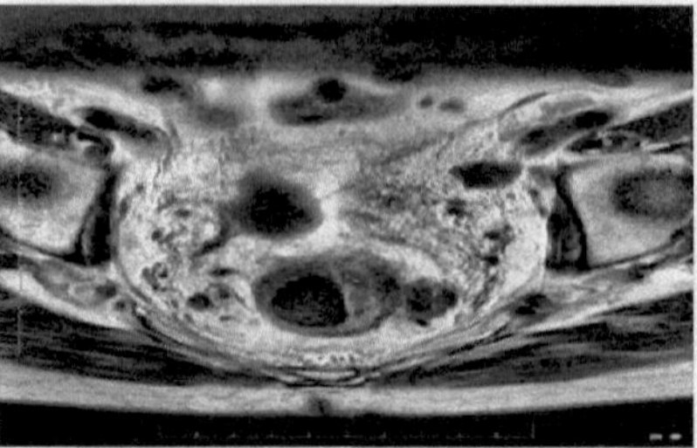

Figure 5.2.2 After preprocessing.

For the analysis of the steps proposed for preprocessing histogram of the original and the image after preprocessing is shown below. It clearly shows how CLAHE redistributed the contrast equally among all the bins in histogram. Ordinary histogram equalization uses the same transformation derived from the image histogram to transform all pixels. This works well when the distribution of pixel values is similar throughout the image. However, when the image contains regions that are significantly lighter or darker than most of the image, the contrast in those regions will not be sufficiently enhanced.

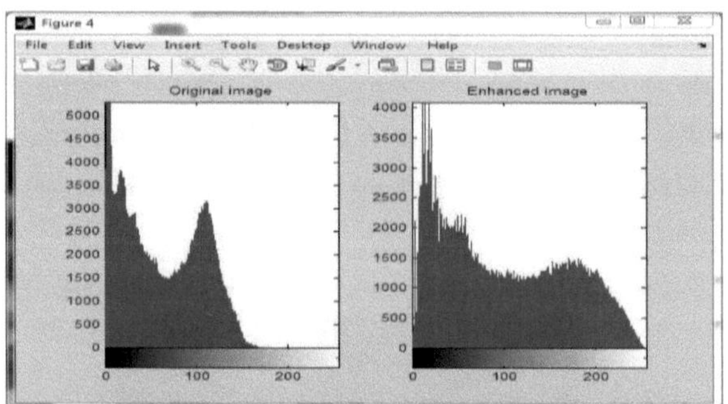

Figure 5.2.3 Histograms of Original and Enhanced.

For this PSNR and AMBE are calculated. PSNR is most commonly used to measure the quality of reconstruction of lossy compression codecs (e.g., for image compression). The signal in this case is the original data, and the noise is the error introduced by compression. When comparing compression codecs, PSNR is an approximation to human perception of reconstruction quality. Although a higher PSNR generally indicates that the reconstruction is of higher quality, in some cases it may not.

Table 5.2.1 AMBE and PSNR values of enhanced images.

Image number	Absolute Mean Brightness Error	Peak Signal Noise Ratio
Image1	37.4788	14.1999
Image2	19.7099	18.4530
Image3	39.1854	14.0689
Image4	46.0584	13.0673
Image5	37.6231	14.2407

5.3 Segmentation

- **Classify the image Using K-Means Clustering**

Clustering is a way to separate groups of objects. Adaptive K-means clustering treats each object as having a location in space. It finds partitions such that objects within each cluster are as close to each other as possible, and as far from objects in other clusters as possible. Using Adaptive k-means to cluster the objects into clusters using the Euclidean distance metric.

- **Label Every Pixel in the Image Using the Results from ADAPTIVE K-MEANS**

For every object in your input, adaptive k-means returns an index corresponding to a cluster. The cluster_center output from adaptive k-means will be used later in the example. Label every pixel in the image with its cluster_id.

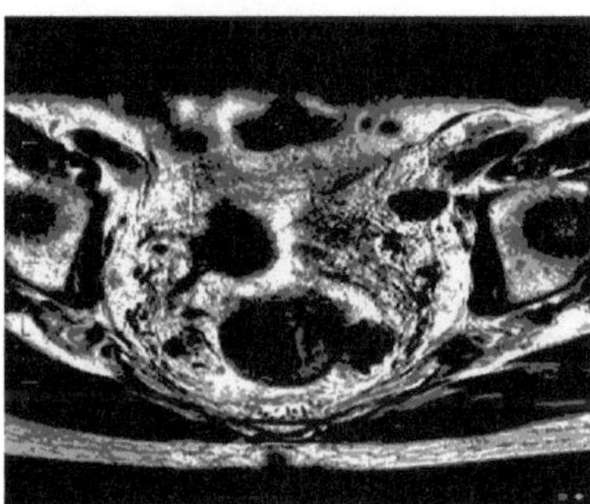

Figure 5.3.1 Image Labelled By Cluster Id.

- **Create Images that Segment the Image**

Using cluster_id, you can separate objects in Image3.jpg, which will result in following images.

Cluster 1:

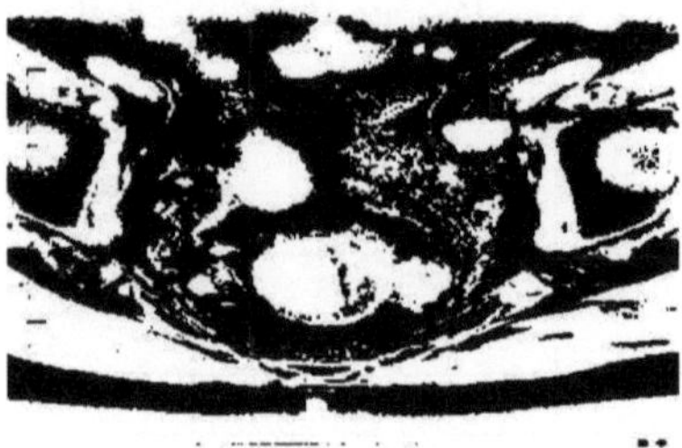

Figure 5.3.2 Objects in cluster 1.

Cluster 2:

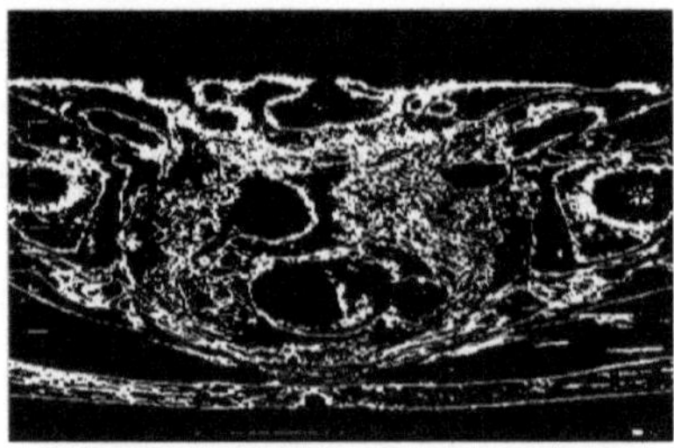

Figure 5.3.3 Objects in cluster 2

Cluster 3:

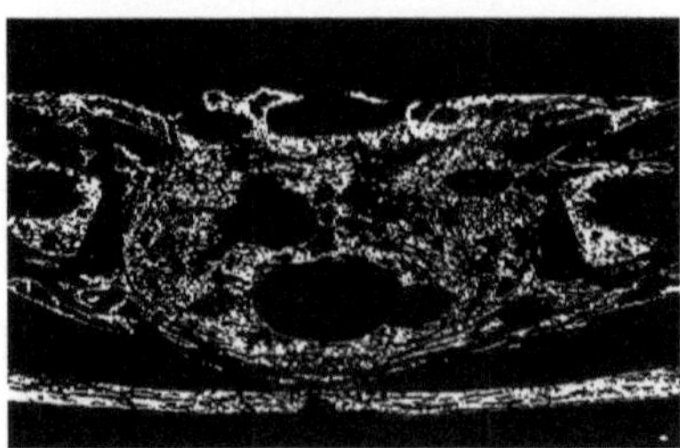

Figure 5.3.4 Objects in cluster 3

Cluster 4:

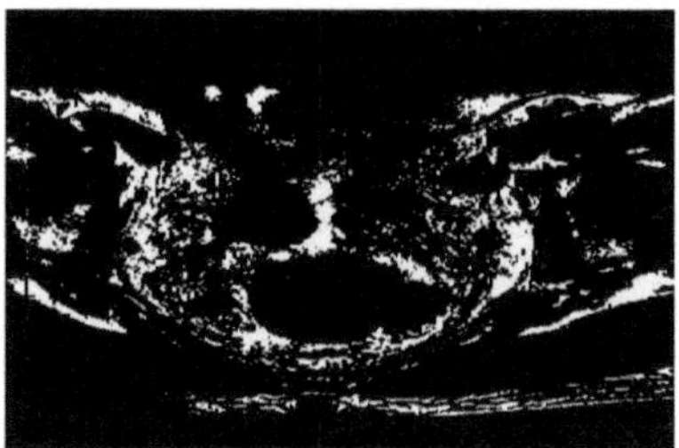

Figure 5.3.5 Objects in cluster 4

After clustering the sub image is selected for post processing. We proposed a framework where the area and mean of the tumor are calculated. These values are useful for future analysis.

5.4 Post processing

Finally the clustered image that is close to the required data is selected and processed further to calculate the area and mean values using image processing techniques. The same process is followed for set of MRI images and the values are exported to excel sheet.

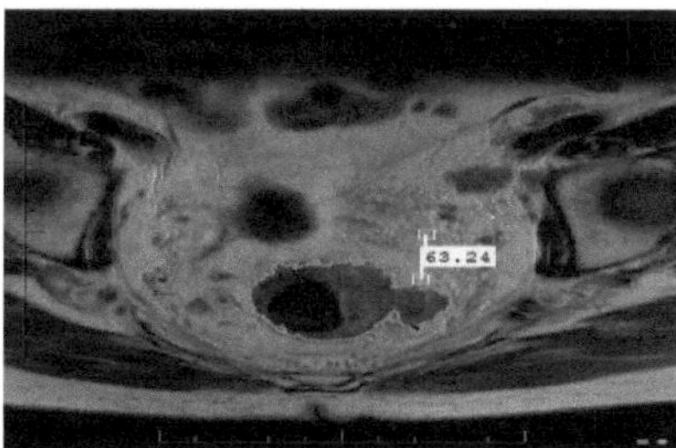

Figure 5.4.1 Object Selected

The boundary is drawn around the tumor area and distance line is used to calculate the distance. The label shows the distance in pixels by default. This value is converted into centimeters by using the equation

$$dcm = dpix *0.02645. \tag{4}$$

where dpix is distance in pixel and dcm is distance in centimeters.

Table 5.4.1 Area and Mean values of enhanced images.

Image number	Area	Mean Values
Image1	33.185	55.591
Image2	33.076	80.171
Image3	29.649	40.964
Image4	16.205	25.314
Image5	31.188	51.801

The distance line of tumor and other parts of the body is calculated by distance line and is shown in the below figure . The line (yellow) shows the distance between the tumor area (red) and mesorectal fascia.

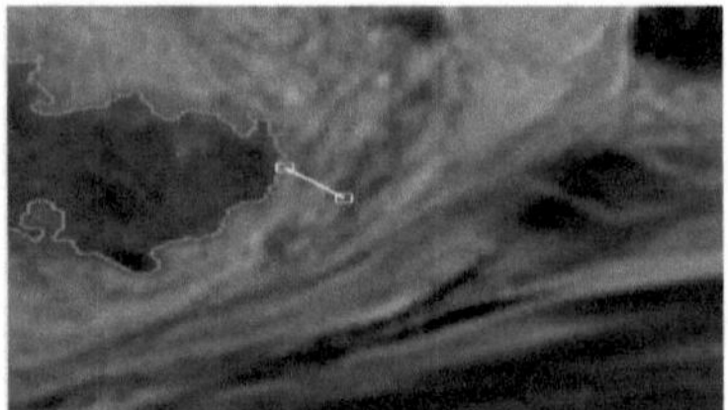

Figure 5.4.2 Distance line.

Finding T3 [17] stage is a crucial one as cancer may spread to other organs of the body. When the tumor enters the mesorectal fat the shortest distance between the tumor and the mesorectal fascia is calculated in the proposed framework. When the tumor reaches the mesorectal fascia the chances of tumor spreading to other organs are more. The distance found to be 0.48. Rectal MRI has the benefits of multiplanar imaging and excellent contrast between tumor and perirectal fat, which helps in tumor detection and its extent for surgical planning and staging especially for low-lying rectal tumors. The mesorectal fascia, which is the border for total mesorectal excision (TME) [18], is clearly seen on MRI.

CHAPTER 6. CONCLUSIONS

Considerable advances have been made in the diagnosis and treatment of rectal cancer over the last decade. Currently, MRI is the only imaging modality that enables prediction of the CRM, as well as preoperative local staging and postoperative assessment. This has helped greatly in determining patient prognosis and planning of surgical therapy.

A Computer Aided Diagnosis (CAD) method is proposed to enhance the input MRI image by image processing techniques. A series of steps are performed on the input image for enhancement of MRI image. The median filter is used for noise reduction, unsharp masking subtracts a blurred version of the image from itself and gives enhanced edges and CLAHE equalizes the contrast and clarity of the image and sharpened image is obtained.

Clustering is done by using adaptive k-means clustering algorithm for the analysis of the colorectal cancer. The selection of tumor part helps in feature extraction like area, mean intensity of the tumor. This is done by the region properties by making boundaries of tumor in the image.

A function called distance line used to calculate the minimum distance from tumor to other parts. This helps the radiologist in staging the cancer.

REFERENCES

[1]. **Avanish P Saklani, Sung Uk Bae, Amy Clayton, and Nam Kyu Kim,** Magnetic resonance imaging in rectal cancer: A surgeon's perspective, *World J Gastroenterol*, 2014; 20(8): 2030–2041.

[2]. **Se Hyung Kim, Jeong Min Lee, Joon-Goo Lee and Philippe A. Lefere** (2007) Computer-Aided Detection of Colonic Polyps at CT Colonography Using a Hessian Matrix–Based Algorithm: Preliminary Study, *Gastrointestinal Imaging*, DOI:10.2214/AJR.07.2072.

[3]. **Abdus Samee and Chelliah Ramachandran Selvasekar** (2011) Current trends in staging rectal cancer, *World Journal of Gastroenterol*, 17(7): 828-834.

[4]. **Christian Klessen, Patrik Rogalla and Matthias Taupitz** (2007) Local staging of rectal cancer: the current role of MRI, *European Radiology*, 17: 379–389.

[5]. **Mussarat Yasmin, Muhammad Sharif, Saleha Masood, Mudassar Raza and Sajjad Mohsin** (2012) Brain Image Enhancement - A Survey, *World Applied Sciences Journal*, 17 (9): 1192-1204.

[6]. **Hanan Saleh S. Ahmed and Md Jan Nordin** (2011) Improving Diagnostic Viewing of Medical Images using Enhancement Algorithms, *Journal of Computer Science*, 7(12): 1831-1838.

[7]. **Michael R. Torkzad, Lars Påhlman and Bengt Glimelius** (2010) Magnetic resonance imaging (MRI) in rectal cancer: a comprehensive review, *European Society of Radiology, Insights Imaging*, 245–267.

[8]. **Giacomo Puppa, Graeme Poston, Per Jess, Guy F Nash, Kenneth Coenegrachts and Axel Stang** (2013) Staging colorectal cancer with the TNM 7th: The presumption of innocence when applying the M category, *World Journal of Gastroenterolgy*, 19(8): 1152-1157.

[9]. **Jiawei Han and Micheline Kamber,** Data Mining: Concepts and Techniques. 3rd Edition, Elsevier, 2011.

[10]. **Gonzalez and Woods,** Digital Image processing. 3rd Edition, Prentice Hall, 2008.

[11]. **Amos Gilat,** MATLAB: An Introduction With Applications.4th Edition, *John Wiley and Sons. INC*, 2010

[12]. **Lim Jae S** (1990), Two-Dimensional Signal and Image Processing, *Englewood Cliffs, NJ, Prentice Hall*, 469-476.

[13]. **Zuiderveld Karel** (1994), Contrast Limited Adaptive Histograph Equalization, *Graphic Gems IV. San Diego: Academic Press Professional,* 474–485.

[14]. **Ekta Thirani** (2013), A Comparative Analysis of Fingerprint Enhancement Technique through Absolute Mean Brightness Error and Entropy, *International Journal on Recent and Innovation Trends in Computing and Communication,* 1(2), 81-84.

[15]. **Kanungo, T.; Mount, D. M.; Netanyahu, N. S.; Piatko, C. D.; Silverman, R.; Wu, A. Y.** (2002). An efficient k-means clustering algorithm: Analysis and implementation. *IEEE Trans. Pattern Analysis and Machine Intelligence* 24: 881–892.

[16]. **Bhagwati Charan Patel and Dr. G.R.Sinha** (2010), An Adaptive K-means Clustering Algorithm for Breast Image Segmentation, *International Journal of Computer Applications*, Volume 10. No.4,.

[17]. **Carolyn C. Compton and Frederick L. Greene** (2004), The Staging of Colorectal Cancer: 2004 and Beyond, *CA Cancer Journal for Clinicians*,54(6),295–308.

[18]. **Ridgway Paul F and Darzi Ara W** (2003), The Role of Total Mesorectal Excision in the Management of Rectal Cancer, *Cancer Control* 10 (3), 205–211.